Everyday Mathematics®

The University of Chicago School Mathematics Project

Skills Link

Cumulative Practice Sets
Student Book

McGraw Hill Wright Group

The McGraw-Hill Companies

Photo Credits

Cover—©Ralph A. Clevenger/CORBIS, cover, *center;* Getty Images, cover, *bottom left;*
©Tom and Dee Ann McCarthy/CORBIS, cover, *right*

Photo Collage—Herman Adler Design

www.WrightGroup.com

 **Wright Group**

Copyright © 2009 by Wright Group/McGraw-Hill.

Printed in the United States of America.

Send all inquiries to:
Wright Group/McGraw-Hill
P.O. Box 812960
Chicago, IL 60681

ISBN 978-0-07-622501-9
MHID 0-07-622501-1

4 5 6 7 8 9 10 MAL 14 13 12 11 10

The McGraw·Hill Companies

Contents

Practice Sets Correlated to Grade 1 Goals

Content	*Everyday Mathematics* Grade 1 Grade-Level Goals	Grade 1 Practice Sets
Number and Numeration		
Rote counting	**Goal 1.** Count on by 1s, 2s, 5s, and 10s past 100 and back by 1s from any number less than 100 with and without number grids, number lines, and calculators.	2, 5, 6, 8, 9, 11, 15, 18, 19, 20, 21, 23, 24, 26, 28, 29, 30, 31, 32, 33, 34, 35, 36, 38, 41, 45, 48, 51, 52, 55, 66, 74, 76, 82, 84
Rational counting	**Goal 2.** Count collections of objects accurately and reliably; estimate the number of objects in a collection.	4, 5, 40, 50, 56, 68
Place value and notation	**Goal 3.** Read, write, and model with manipulatives whole numbers up to 1,000; identify places in such numbers and the values of the digits in those places.	1, 4, 6, 10, 11, 68, 73, 74, 81, 88
Meanings and uses of fractions	**Goal 4.** Use manipulatives and drawings to model halves, thirds, and fourths as equal parts of a region or a collection; describe the model.	71, 72, 73, 79, 80, 81
Number theory	**Goal 5.** Use manipulatives to identify and model odd and even numbers.	18, 20, 35, 48, 53
Equivalent names for whole numbers	**Goal 6.** Use manipulatives, drawings, tally marks, and numerical expressions involving addition and subtraction of 1- or 2-digit numbers to give equivalent names for whole numbers up to 100.	4, 5, 7, 12, 13, 46, 49, 54, 60, 62, 67, 80
Comparing and ordering numbers	**Goal 7.** Compare and order whole numbers up to 1,000.	3, 7, 9, 15, 22, 33, 39, 42, 43, 44, 81, 88
Operations and Computation		
Addition and subtraction facts	**Goal 1.** Demonstrate proficiency with +/– 0, +/– 1, doubles, and sum-equals-ten addition and subtraction facts such as 6 + 4 = 10 and 10 – 7 = 3.	10, 37, 46, 49, 50, 51, 53, 57, 62, 63, 67, 68, 70, 72, 75, 88
Addition and subtraction procedures	**Goal 2.** Use manipulatives, number grids, tally marks, mental arithmetic, and calculators to solve problems involving the addition and subtraction of 1-digit whole numbers with 1- or 2-digit whole numbers; calculate and compare the values of combinations of coins.	8, 10, 14, 15, 16, 17, 22, 36, 40, 43, 47, 51, 55, 56, 60, 63, 65, 69, 70, 75, 77, 78, 81, 82, 84, 85, 87
Computational estimation	**Goal 3.** Estimate reasonableness of answers to basic fact problems (e.g., Will 7 + 8 be more or less than 10?).	21, 50, 62, 76
Models for the operations	**Goal 4.** Identify change-to-more, change-to-less, comparison, and parts-and-total situations.	16, 28, 32

Content	Everyday Mathematics Grade 1 Grade-Level Goals	Grade 1 Practice Sets
Data and Chance		
Data collection and representation	**Goal 1.** Collect and organize data to create tally charts, tables, bar graphs, and line plots.	25, 86
Data analysis	**Goal 2.** Use graphs to ask and answer simple questions and draw conclusions; find the maximum and minimum of a data set.	16, 27, 32, 40, 46, 58, 82, 86
Qualitative probability	**Goal 3.** Describe events using *certain, likely, unlikely, impossible* and other basic probability terms.	13, 14, 57, 87
Measurement and Reference Frames		
Length, weight, and angles	**Goal 1.** Use nonstandard tools and techniques to estimate and compare weight and length; measure length with standard measuring tools.	30, 31, 39, 41, 42, 44, 52, 77, 85, 86
Money	**Goal 2.** Know and compare the value of pennies, nickels, dimes, quarters, and dollar bills; make exchanges between coins.	16, 17, 23, 27, 38, 45, 51, 65, 66, 67, 70, 83, 84
Temperature	**Goal 3.** Identify a thermometer as a tool for measuring temperature; read temperatures on Fahrenheit and Celsius thermometers to the nearest 10°.	7, 23, 24, 29, 87
Time	**Goal 4.** Use a calendar to identify days, weeks, months, and dates; tell and show time to the nearest half and quarter hour on an analog clock.	11, 12, 17, 21, 25, 34, 54, 56, 57, 58, 61, 66
Geometry		
Plane and solid figures	**Goal 1.** Identify and describe plane and solid figures including circles, triangles, squares, rectangles, spheres, cylinders, rectangular prisms, pyramids, cones, and cubes.	59, 60, 61, 62, 63, 64
Transformations and symmetry	**Goal 2.** Identify shapes having line symmetry; complete line-symmetric shapes or designs.	65, 68, 71, 78, 83
Patterns, Functions, and Algebra		
Patterns and functions	**Goal 1.** Extend, describe, and create numeric, visual, and concrete patterns; solve problems involving function machines, "What's My Rule?" tables, and Frames-and-Arrows diagrams.	18, 24, 26, 47, 48, 49, 54, 59, 64, 75, 76, 78, 79
Algebraic notation and solving number sentences	**Goal 2.** Read, write, and explain expressions and number sentences using the symbols +, −, and = and the symbols > and < with cues; solve equations involving addition and subtraction.	42, 43, 63, 83, 84
Properties of arithmetic operations	**Goal 3.** Apply the Commutative Property of Addition and the Additive Identity to basic addition fact problems.	45, 51, 60, 70, 88

Grade K Review: Number and Numeration

1. Write the number cards in order
   from least to greatest.

| 14 | 9 | 17 | 5 | 10 |

| | | | | |

2. Circle half the number of kittens.

3. How many flowers? _____

4. Take away 4 cars. How many are left? _____

Grade K Review: Number and Numeration

5. Count on.

0, _____, _____, 3, _____, _____, _____, 7, _____, _____, 10

6. Count on by 2s.

0, _____, 4, _____, _____, 10, _____, 14, _____, _____, 20

7. Count back by 10s.

100, _____, 80, _____, 60, _____, 40, _____, 20, _____, 0

Write the number before and after.

8. _____, 12, _____   **9.** _____, 7, _____

10. _____, 25, _____   **11.** _____, 40, _____

Write the numbers from smallest to largest.

    8   3   13   0   20

12. _____, _____, _____, _____, _____

Draw a picture to show each number.

13. 10   **14.** 6

Grade K Review: Operations and Computation

1. How many marbles in all? Write the number sentence.

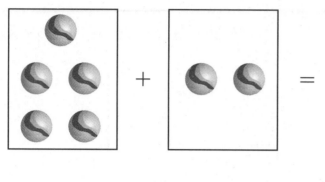

____ + ____ = ____

2. Draw 4 more faces. Write the number sentence.

3 + ____ = ____

3. How many rabbits are left? Write the number sentence.

8 − ____ = ____

4. How many squirrels in all? Write the number sentence.

5 + ____ = ____

Grade K Review: Operations and Computation

Use the number line. Write the missing number.

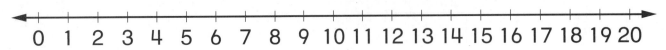

5.

Total	
Part	**Part**
7	5

6.

Total	
Part	**Part**
3	10

7.

Start
12

Change

End

7 more

8.

Start
9

Change

End

3 less

Write the sum.

9. + = _____

10.  +  = _____

Solve.

11. There are 12 ladybugs. There are 4 flowers. Draw an equal number of ladybugs on each flower.

Grade K Review: Data and Chance

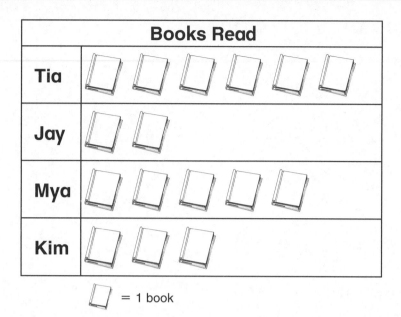

Books Read

Tia	📖 📖 📖 📖 📖 📖
Jay	📖 📖
Mya	📖 📖 📖 📖 📖
Kim	📖 📖 📖

☐ = 1 book

1. Who read the most books? _____

   How many books were read? _____

2. Who read the fewest books? _____

   How many books were read? _____

3. How many books did the children read in all? _____

Look at the bag of marbles.

4. How likely are you to pick
   a white marble without looking?

   unlikely   likely   certain

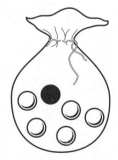

5. How likely are you to pick a black marble without looking?

   unlikely   likely   certain

Grade K Review: Measurement and Reference Frames

Write the time on each clock.

1.

_____ o'clock

2.

_____ o'clock

Write your answer.

3. 1 minute = _____ seconds

4. 1 hour = _____ minutes

Look at the coins.

5. How many pennies are there? _____

6. How many nickels are there? _____

7. How many dimes are there? _____

8. How many quarters are there? _____

Grade K Review: Measurement and Reference Frames

**Draw lines to match each picture
to a temperature.**

9.

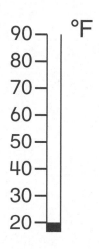

°F
90
80
70
60
50
40
30
20

10.

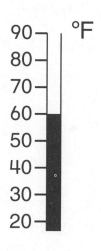

°F
90
80
70
60
50
40
30
20

11.

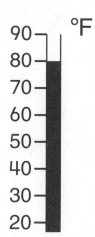

°F
90
80
70
60
50
40
30
20

Grade K Review: Geometry

Write the name of each shape.

triangle   circle   rectangle   square

1. 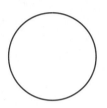 _____

2.  _____

3. _____

4.  _____

5. Look at the first shape. Circle the picture that shows the whole shape.

  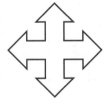 

6. What looks like a cube? Circle the shape.

 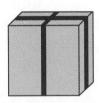

7. What looks like a sphere? Circle the shape.

 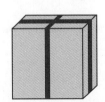

Grade K Review: Patterns, Functions, and Algebra

Write the rule. Write the missing *out* numbers.

1.

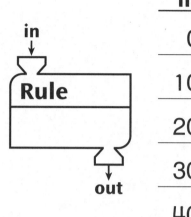

in	out
0	5
10	15
20	25
30	
40	

2.

in	out
10	8
9	7
8	6
7	
6	

Add.

3. 6 + 5 = _____

4. 8 + 4 = _____

5. 9 + 2 = _____

6. 10 + 6 = _____

Finish the pattern.

7. ◯, ▢, △, ◯, ▢, △, _____, _____, _____

Write the missing number.

8. 6 + ▢ = 9

9. ▢ + 4 = 10

10. 2 − ▢ = 2

11. 5 + ▢ = 5

Practice Set 1

1. Write the number.

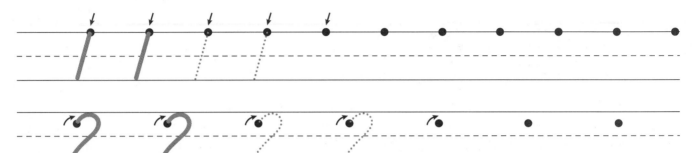

2. Draw a picture of 1 thing.	**3.** Draw a picture of 2 things.
4. Draw circles to show 5.	**5.** Draw sticks to show 10.

Practice Set 2

1. Draw 1 more.

2. Show 1 less. Make an X.

3. Count by 1s. Connect the dots.

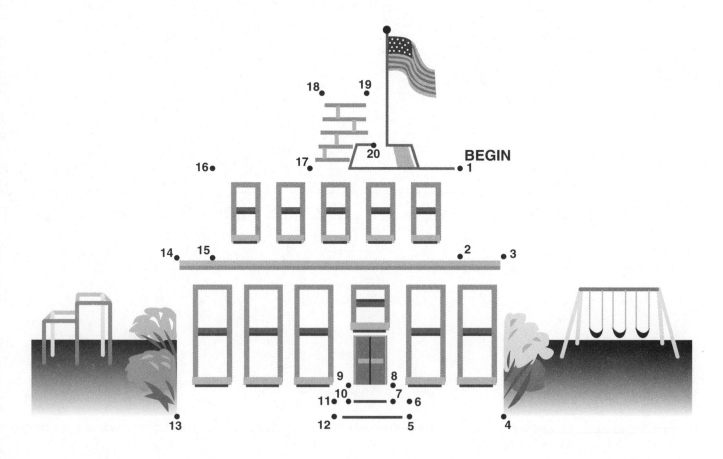

Use with or after Lesson 1•6.

Practice Set ⟨ 3 ⟩

Circle the larger number in each pair.

1.

2.

Circle the mystery number below each number line.

3.

1   2   3   ☐   5   6   7   8   9

2   4   7

4.

1   2   3   4   5   6   7   ☐   9

2   5   8

5.

1   2   ☐   4   5   6   7   8   9

3   6   8

Practice Set ◆ 4

Make tallies to show each number.

Example 7 _HHT //_____   **1.** 5 _____

2. 4 _____   **3.** 9 _____

4. 11 _____   **5.** 16 _____

6. Write the number.

7. How many balls?

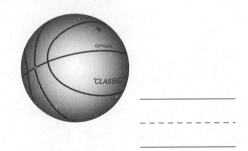

_____
- - - - - - - - - -
_____

8. How many cars?

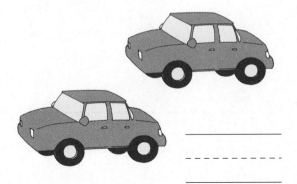

_____
- - - - - - - - - -
_____

Practice Set 5

1. Fill in the missing numbers.
    Next, circle the 10 and the 15.
    Then cross out the 7.

February						
Sunday	Monday	Tuesday	Wednesday	Thursday	Friday	Saturday
1	___	___	___	___	___	7
8	9	10	11	12	13	14
15	16	17	18	19	20	21
22	23	24	25	26	27	28

2. Write the number.

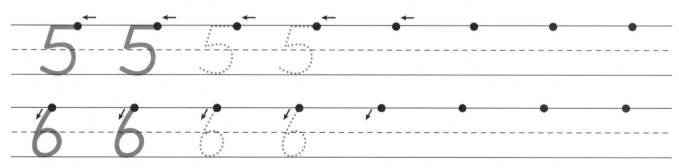

3. Draw squares to show 4.

4. Draw a picture of 3 things.

Practice Set 6

Make the shape. Use your Pattern-Block Template.

1.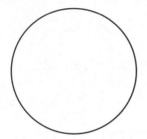

2.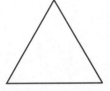

3.

4. You roll   .

Circle that many pennies.

5. You roll   .

Circle that many pennies.

Use with or after Lesson 1·11.

Practice Set 7

Match each picture to a temperature.

1.

2.

3.

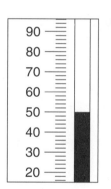

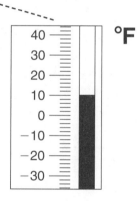

---

4. Show how old you are. Use tally marks. _____

5. Show how many pencils you have.
Use tally marks. _____

Circle the larger number.

6. 5   8

7. 21   36

Practice Set ⟨8⟩

Solve the number stories.
Use the pictures to help you.

1. You have 5 pennies. Your friend gives you 4 pennies.

How many pennies do you have now? _____

Ⓟ Ⓟ Ⓟ Ⓟ Ⓟ   Ⓟ Ⓟ Ⓟ Ⓟ

2. You have 10 pennies. You lose 2 of them.

How many pennies do you have now? _____

Ⓟ Ⓟ Ⓟ Ⓟ Ⓟ Ⓟ Ⓟ Ⓟ Ⓟ Ⓟ

3. Joey earned $1.00 a day for three days.

How much did he earn? _____

Day 1   **Day 2**   **Day 3**

Write the number.

4. What number comes after 7? _____

5. What number comes before 8? _____

6. What number comes after 5? _____

7. What number is 1 less than 11? _____

Practice Set 9

									0
1	2	3	4	5	6	7	8	9	10
11	12	13	14	15	16	17	18	19	20
21	22	23	24	25	26	27	28	29	30
31	32	33	34	35	36	37	38	39	40
41	42	43	44	45	46	47	48	49	50
51	52	53	54	55	56	57	58	59	60
61	62	63	64	65	66	67	68	69	70
71	72	73	74	75	76	77	78	79	80
81	82	83	84	85	86	87	88	89	90
91	92	93	94	95	96	97	98	99	100
101	102	103	104	105	106	107	108	109	110

1. Start at 7. Count up 8 spaces.
   Color the box **red.**

2. Start at 45. Count up 9 spaces.
   Color the box **yellow.**

3. Start at 18. Count back 7 spaces.
   Color the box **purple.**

4. Start at 102. Count back 6 spaces.
   Color the box **green.**

5. Start at 103. Count back 10 spaces.
   Color the box **blue.**

Practice Set 10

Circle each pair that shows 10 pennies.

1.

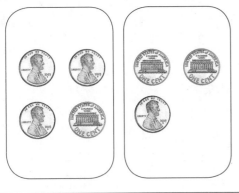

2.

3.

4.

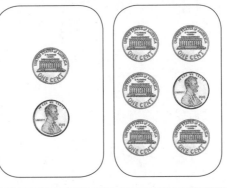

5.

6.

7. Write the number.

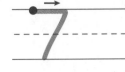

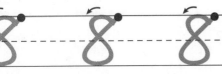

Use with or after Lesson 2•3.

Practice Set 11

1. Write the missing numbers on the clock.

2. What time does the clock show?

| 9 o'clock | between 9 o'clock and 10 o'clock | 10 o'clock |

3. Write the number.

 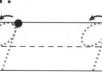

   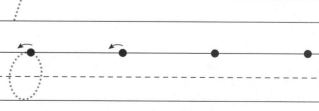

Practice Set 12

What time is it?

1.

_____ o'clock

2.

_____ o'clock

3.

_____ o'clock

Draw the hour hand.

4.

3 o'clock

5.

8 o'clock

6.

11 o'clock

7. Make tallies to show 9.

8. Make tallies to show 10.

9. Make tallies to show 13.

10. Make tallies to show 6.

Practice Set 13

Match the dominoes.

1.

2.

3.

How many dots in all?

4.  _____

5.  _____

6. What color marble is Marla
   most likely to pick without looking?

   _____

7. How likely is it that Marla will
   pick a striped marble?

   certain   likely   unlikely   impossible

Practice Set 14

**Show each amount using fewer coins.
Draw Ⓝs and Ⓟs.**

Example

Ⓝ   Ⓟ

1.

2.

3. How likely is it that the spinner will land on △ ?
       certain   likely   unlikely   impossible

4. How likely is it that the spinner will land on ⬠ ?
       certain   likely   unlikely   impossible

5. How likely is it that the spinner will land on ☐ ?
       certain   likely   unlikely   impossible

     Use with or after Lesson 2•9.

Name _____ Date _____ Time _____

Practice Set 15

Count by 5s. Tell how much.

1.

 15¢

2.

3.

4.

How much money?

5.   _____¢

6.   _____¢

Circle the larger number.

7. 11 12   **8.** 7 15   **9.** 18 17

Use with or after Lesson 2·10.

Practice Set 16

MRB
39 88
116

Complete each diagram. Then complete
the number model.

1.
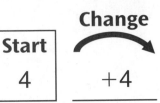

Change

Start		End
4	+4	

Number model:

4 + 4 = _____

2.
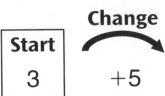

Change

Start		End
3	+5	

Number model:

3 + 5 = _____

Use the tally chart below to answer the questions.

Favorite Colors	
Colors	**Tallies**
Red	~~||||~~ ||||
Blue	~~||||~~ ~~||||~~
Other	||||

3. How many people picked **red** as a favorite color? _____

4. How many picked **blue** as a favorite color? _____

5. How many picked a color **other** than red or blue? _____

How much money?

6. Ⓝ Ⓝ Ⓝ Ⓟ Ⓟ Ⓟ Ⓟ _____ ¢

7. Ⓝ Ⓝ Ⓟ Ⓟ Ⓟ Ⓟ Ⓟ _____ ¢

8. Ⓝ Ⓟ Ⓟ Ⓟ _____ ¢

9. Ⓝ Ⓝ Ⓟ _____ ¢

Use with or after Lesson 2•11.

Practice Set 17

1. Cross out 3 pennies. How much money is left?

_____ ¢

2. Cross out 2 nickels. How much money is left?

_____ ¢

Draw **s and** **s to show how much.**

3. 7 cents

---

4. 12 cents

What time is it?

5.

_____ : _____

6.

_____ : _____

Practice Set 18

Circle all the pairs. Then circle *even* or *odd*.

1.

even   odd

2. ☆ ☆ ☆ ☆ ☆ ☆
☆ ☆ ☆ ☆ ☆ ☆

even   odd

Draw the next shape.
Use your Pattern-Block Template.

3.

4.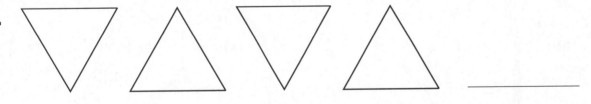

5. Count by 2s.

__2__ , __4__ , __6__ , _____ , _____ , _____ , _____ , _____

Practice Set 19

1. Count by 3s. Put an **X** over each number
   you count.

									0
1	2	X̶	4	5	X̶	7	8	X̶	10
11	12	13	14	15	16	17	18	19	20
21	22	23	24	25	26	27	28	29	30
31	32	33	34	35	36	37	38	39	40
41	42	43	44	45	46	47	48	49	50
51	52	53	54	55	56	57	58	59	60
61	62	63	64	65	66	67	68	69	70
71	72	73	74	75	76	77	78	79	80
81	82	83	84	85	86	87	88	89	90
91	92	93	94	95	96	97	98	99	100
101	102	103	104	105	106	107	108	109	110

Write the missing numbers.

2. ← —— 8 —— __ —— __ —— 11 —— __ —— __ —— __ →

3. ← —— __ —— 22 —— __ —— __ —— 26 —— __ →

Practice Set 20

Write the number of dots.
Then write *even* or *odd*.

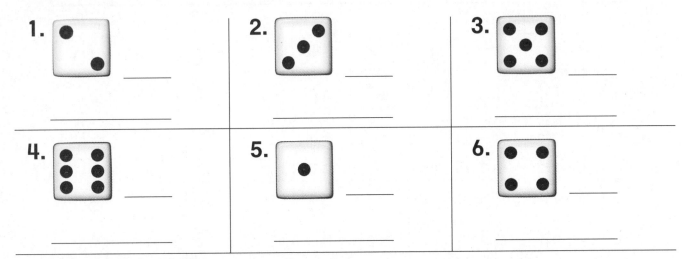

1. _____

_____

2. _____

_____

3. _____

_____

4. _____

_____

5. _____

_____

6. _____

_____

7. **Writing/Reasoning** Is 8 is an even number or an odd number? Explain your answer by using words, numbers, or pictures.

8. Count by 10s.

_____, _____, 50, 60, _____, _____, 90, _____

9. Count by 1s.

_____, _____, 26, 27, _____, _____, _____, 31

10. Count by 5s.

20, _____, _____, 35, 40, _____, _____, _____

34

Practice Set 21

1. Count by 2s. Show your counts.

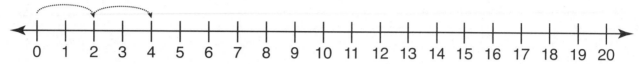

0 1 2 3 4 5 6 7 8 9 10 11 12 13 14 15 16 17 18 19 20

2. Count back by 1s. Show your counts.

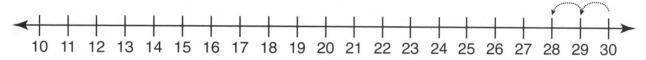

10 11 12 13 14 15 16 17 18 19 20 21 22 23 24 25 26 27 28 29 30

Record the time.

3.

about _____ o'clock

4.

between _____:00 and _____:00

5. Will 10 − 7 be more or less than 5? _____

6. Will 8 + 5 be more or less than 10? _____

7. Will 9 + 3 be more or less than 10? _____

8. Count by 3s.

_6_ , _9_ , _____ , _____ , _18_ , _____ , _____ , _27_

Practice Set 22

Use the number line to help you solve these problems.

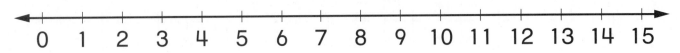

0 1 2 3 4 5 6 7 8 9 10 11 12 13 14 15

1. Start at 5. Count up 2 hops. Where do you
   end up? ____

$$5 + 2 = \text{\_\_\_\_}$$

2. Start at 11. Count back 4 hops. Where do you
   end up? ____

$$11 - 4 = \text{\_\_\_\_}$$

3. Start at 15. Count back 3 hops. Where do you
   end up? ____

$$15 - 3 = \text{\_\_\_\_}$$

Circle the larger number.

4. 34   15   5. 11   12

6. 0   1   7. 21   12

8. ✎ **Writing/Reasoning** Explain how you found
   your answer to Problem 6.

   _____

   _____

   _____

Practice Set 23

Color the thermometer to show the temperature.

1. 80°F

°F

100
90
80
70
60
50

2. 30°F

°F

40
30
20
10
0
−10

3. 100°F

°F

140
130
120
110
100
90

How much money?

4.  $0.____ or ____¢

5.  $0.____ or ____¢

6.  $0.____ or ____¢

7. $0.____ or ____¢

8. $0.____ or ____¢

Practice Set 24

Fill in the frames.

Example

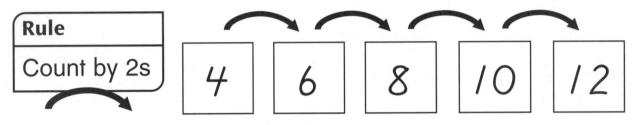

Rule	
Count by 2s	4 6 8 10 12

1.

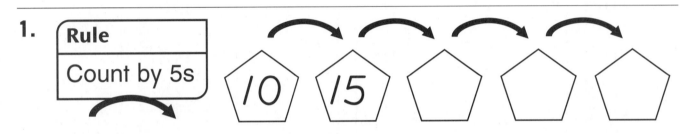

Rule	
Count by 5s	10 15

2.

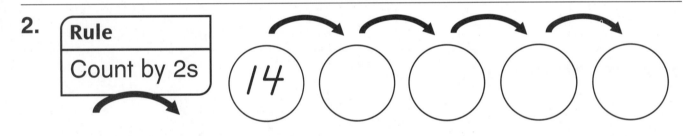

Rule	
Count by 2s	14

Record the temperature.

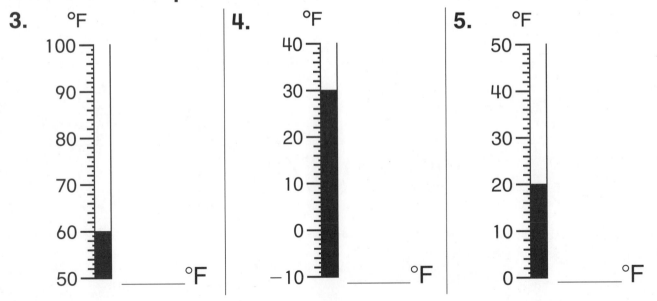

3. °F
100 — 90 — 80 — 70 — 60 — 50

_____ °F

4. °F
40 — 30 — 20 — 10 — 0 — −10

_____ °F

5. °F
50 — 40 — 30 — 20 — 10 — 0

_____ °F

Use with or after Lesson 3•8.

Practice Set 25

MRB
40 82 83

Count the shapes.
Use tally marks to complete the chart.

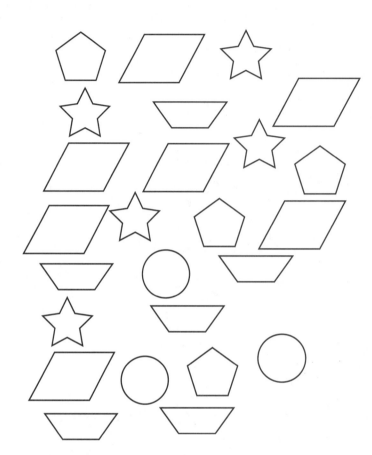

Tally Chart	
Shape	**How Many?**
1. ☆	
2. ⬠	
3. ▱	
4. ▱	
5. ◯	

Tell the time.

6.

half-past _____ o'clock

7.

half-past _____ o'clock

Practice Set 26

1. Use your calculator. Count up by 2s.

Press ⓄN/C ② ⊕ ② ⊜ ⊜ ⊜ ⊜

_2_ , _4_ , _6_ , _8_ , ____ , ____ , ____ , ____ ,

_18_ , ____ , ____ , ____ , ____ , ____ , ____ , _32_

2. Use your calculator. Count back by 2s.

Press ⓄN/C ③ ④ ⊖ ② ⊜ ⊜ ⊜ ⊜

_34_ , _32_ , _30_ , _28_ , ____ , ____ , ____ , ____ ,

_18_ , ____ , ____ , ____ , ____ , ____ , ____ , ____ ,

Fill in the rule box. Complete the frames.

3.

4.

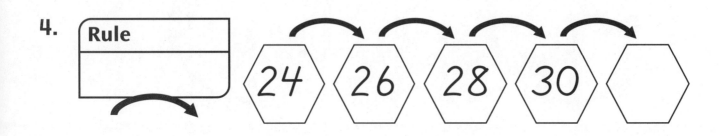

         Use with or after Lesson 3·10.

Practice Set 27

How much money?

1. 1 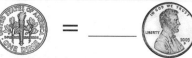 = _____

2. 20 = _____

3. 2  = _____

4. 2   = _____

5. $0._____ or _____¢

6. $0._____ or _____¢

7. $0._____ or _____¢

How many children have ...

8. 5 people in their family?

_____ children

9. 4 people in their family?

_____ children

10. 2 or 3 people in their family?

_____ children

People in My Family	
Number of People	**Frequency**
2	///
3	/////
4	///// ///// ///// /
5	///// ///// //
6	//

Practice Set 28

Write 3 numbers for each domino.

1.

Total

Part	Part
3	2

2.

Total

Part	Part

3.

Total

Part	Part

4.

Total

Part	Part

5.

Total

Part	Part

6.

Total

Part	Part

Write the numbers before and after.

7. _____, 38, _____

8. _____, 92, _____

9. _____, 79, _____

10. _____, 100, _____

11. _____, 15, _____

12. _____, 55, _____

13. _____, 27, _____

14. _____, 80, _____

Use with or after Lesson 3·14.

Practice Set 29

Write the °F temperature.

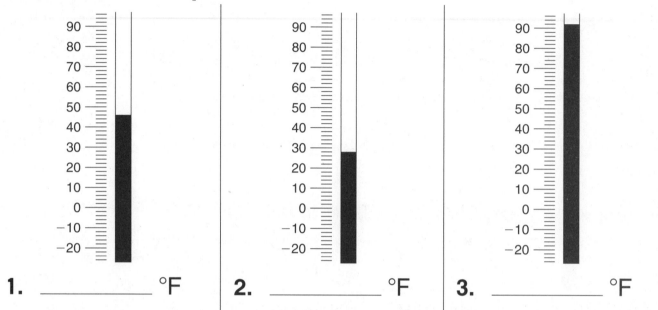

1. _____ °F   2. _____ °F   3. _____ °F

4. Fill in the missing numbers. Count by 3s.
Circle each number you count.

									0
1	2	③	4	5	⑥	7	8	⑨	10
11		13	14		16	17		19	20
	22	23		25	26		28	29	
31	32		34	35		37	38		40
41		43	44		46	47		49	50
	52	53		55	56		58	59	
61	62		64	65		67	68		70
71		73	74		76	77		79	80
	82	83		85	86		88	89	
91	92		94	95		97	98		100
101		103	104		106	107		109	110

Practice Set 30

About how long?

1.

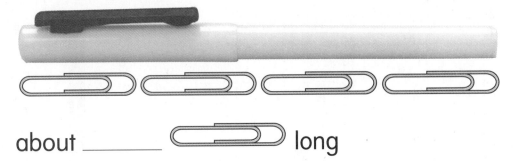

about _____ ⊂⊃ long

2.

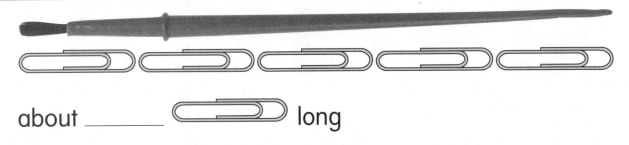

about _____ ⊂⊃ long

3.

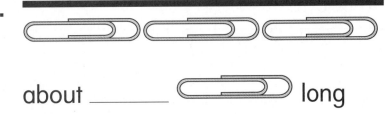

about _____ ⊂⊃ long

4. 📝 **Writing/Reasoning** Which object is longest? Explain your answer by using words, numbers, or pictures.

Write the numbers before and after.

5. _____, *19*, _____

6. _____, *43*, _____

7. _____, *30*, _____

8. _____, *27*, _____

Practice Set 31

MRB
6 65

Measure each object.

1.

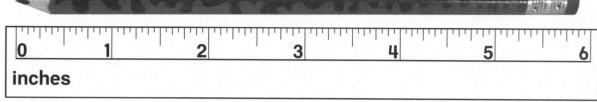

about _____ inches long

2.

about _____ inches long

3.

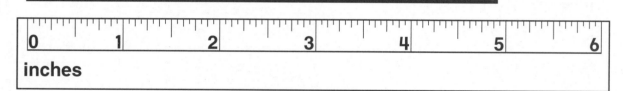

about _____ inches long

Count by 10s. Show your counts.

4.

0 1 2 3 4 5 6 7 8 9 10 11 12 13 14 15 16 17 18 19 20 21 22 23 24 25 26 27 28 29 30 31 32 33

5.

14 15 16 17 18 19 20 21 22 23 24 25 26 27 28 29 30 31 32 33 34 35 36 37 38 39 40 41 42 43 44 45 46 47

Practice Set 32

MRB
7 44
108 109

Find the missing numbers.

1.

Total	
Part	**Part**
8	3

2.

Total	
Part	**Part**
7	6

3.

Total	
Part	**Part**
2	8

4.

Total	
13	
Part	**Part**
8	

5.

Total	
15	
Part	**Part**
	7

6.

Total	
12	
Part	**Part**
5	

Use the bar graph to answer the questions.

7. How many children like grapes?

_____

8. Which fruit did children like most?

_____

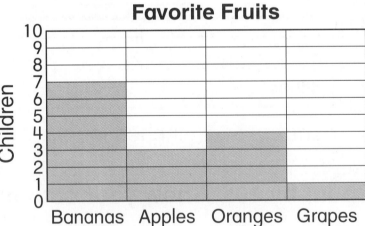

9. **Writing/Reasoning** Explain how you found your answer to Problem 8.

_____

_____

Practice Set 33

Children's Heights

| Sally | Ken | Drew | Kendra | Lorna |
| 41 in. | 46 in. | 46 in. | 46 in. | 49 in. |

1. How tall is Sally?

   Sally is ____ inches tall.

2. Who is tallest?

   _____ is tallest.

3. How many children are 46 inches tall?

   ____ children are 46 inches tall.

Write the numbers before and after.

4. ____, *17*, ____

5. ____, *30*, ____

Fill in the missing numbers. Count down.

6. *10*, ____, ____, *7*, ____, ____, ____, ____, ____, ____

7. *18*, ____, *16*, ____, ____, ____, ____, ____, ____, ____

8. *33*, ____, ____, *30*, ____, ____, ____, ____, ____, ____

Practice Set 34

Record the time.

1.

_____ o'clock

2.

half-past_____ o'clock

3.

quarter-past _____ o'clock

4.

quarter-to _____ o'clock

Count by 10s.

5. _10_, _20_, ____, ____, ____, ____, ____, ____, ____, ____

6. _40_, ____, _60_, ____, ____, ____, ____, ____, ____, ____

7. _50_, ____, ____, ____, _90_, ____, ____, ____, ____, ____

8. _0_, ____, ____, ____, _40_, ____, ____, ____, ____, ____

Use with or after Lesson 4•8.

Practice Set 35

									0
1	2	3	4	5	6	✗	8	9	10
11	✗	13	14	15	16		18	19	20
21		23	24	25	26		28	29	30
31		33	34	35	36		38	39	40
41		43	44	45	46		48	49	50
									60
61	62	63		65				69	70
71	72		74	75	76	77	78	79	
81	82	83	84		86		88	89	90
	92	93	94	95				99	100
		103	104			107	108	109	
	112	113			116	117		119	120

1. Fill in the missing numbers on the grid.

2. Start at 7. Count by 5s.
   Put an **X** over each number you count.

3. Find three **odd** numbers. Draw circles
   around them.

Count down. Fill in the missing numbers.

4. _75_, ____, ____, _72_, ____, ____, _69_, ____, ____

Practice Set 36

Find the sum of each pair of dice.

Example

___2___ + ___3___ = ___5___

1.

_____ + _____ = _____

2.

_____ + _____ = _____

3.

_____ + _____ = _____

4.
+

5.
+

6.
+

Write the missing numbers.

7.

___−1___ ___0___ _____ _____ _____ _____ ___6___

8.

___8___ _____ ___10___ _____ _____ _____ _____

Practice Set 37

Find each sum. Then color.

Colors	
6	yellow
7	red
8	green
9	blue
10	orange

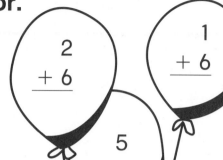

$2 + 6$

$1 + 6$

$5 + 4$

$9 + 0$

$4 + 5$

$2 + 7$

$8 + 1$

$4 + 4$

$5 + 2$

_____ $= 0 + 8$

$6 + 3 =$ _____

$4 + 3$

$5 + 5$

$2 + 6 =$ _____

$5 + 3$

$3 + 3 =$ _____

$1 + 5 =$ _____

$1 + 7$

Practice Set 38

1. Use your calculator. Count up by 10s.

   Press (ON/C)(1)(0)(+)(1)(0)(=)(=)(=)

   _10_, _20_, _30_, _40_, _____, _____, _____, _____, _____,

   _____, _____, _____, _____, _____, _____, _____, _____, _____,

   _____, _____, _____, _____, _____, _____, _____, _____, _____

Show how much. Use the fewest coins.

Draw Ⓠs, Ⓓs, Ⓝs, or Ⓟs.

2. $0.28 or 28¢

3. $0.41 or 41¢

4. Show 35 cents using two different sets of coins.

Practice Set 39

Write <, >, or =.

> < means *is less than*
> > means *is greater than*
> = means *is equal to*

1. 12 ☐ 19   **2.** 43 ☐ 34

3. 115 ☐ 115   **4.** 58 ☐ 68

5. 124 ☐ 224   **6.** 300 ☐ 299

Mark the line segment to show the length.

Example 3 inches

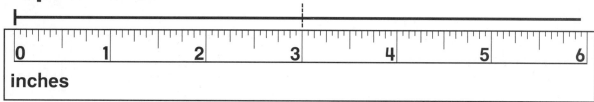

7. 5 inches

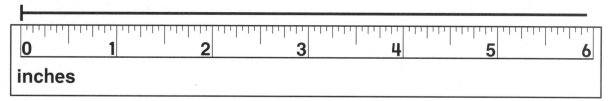

8. 2 inches

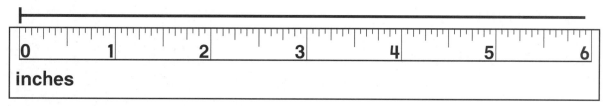

Practice Set 40

How many units?

1.

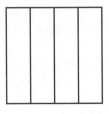

_____ units

2.

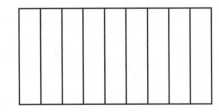

_____ units

3.

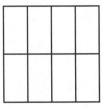

_____ units

4.

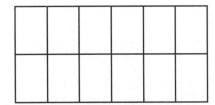

_____ units

Answer each question.

5. How many children chose a dog?

_____ children

6. Did more children choose a horse or a monkey?

_____

How many more?

_____ more children

Favorite Animal	
Horse	///
Elephant	✝✝✝
Dog	✝✝✝ ✝✝✝ ✝✝✝ /
Monkey	✝✝✝ ✝✝✝ //
Cat	✝✝✝ ✝✝✝ //
Turtle	//

54

Practice Set 41

Write the total weight.

1.

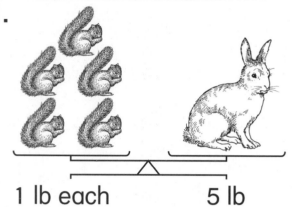

1 lb each   5 lb

_____ pounds

2.

300 lb   50 lb

_____ pounds

Count by 10s.

3. _70_, ____, ____, _100_, ____, ____, ____, ____, ____

4. _160_, ____, _180_, ____, ____, ____, ____, ____, ____

5. _36_, _46_, ____, ____, ____, ____, ____, _106_, ____

6. _92_, _102_, ____, ____, ____, ____, ____, ____, ____

7. ____, _34_, ____, ____, _64_, ____, ____, ____, ____

8. ____, ____, _29_, ____, ____, _59_, ____, ____, ____

9. ____, ____, ____, _43_, ____, ____, _73_, ____, ____

10. _157_, _167_, ____, ____, ____, _207_, ____, ____, ____

Practice Set 42

Write <, >, or = in each circle.

1.

Beaver
56 lb

Girl
50 lb

2.

Five Squirrels
1 lb each

Rabbit
5 lb

3.

Black Bear
300 lb

Two Children
50 lb each

4. ✎ **Writing/Reasoning** How did you solve Problem 3? Explain your answer by using words, pictures, or numbers.

Practice Set 43

How many more?

Example

_____3_____ more pennies

1.

_____ more pennies

2.

_____ more pennies

Write <, >, or =.

3. 10 ☐ 20   **4.** 17 ☐ 17

5. 59 ☐ 69   **6.** 56 ☐ 57

> < means *is less than*
> > means *is greater than*
> = means *is equal to*

Make these true.

Example ☐14☐ < ☐20☐   **7.** ☐18☐ > ☐

8. ☐30☐ > ☐   **9.** ☐ < ☐14☐

10. ☐103☐ > ☐   **11.** ☐9☐ = ☐ + ☐

Practice Set 44

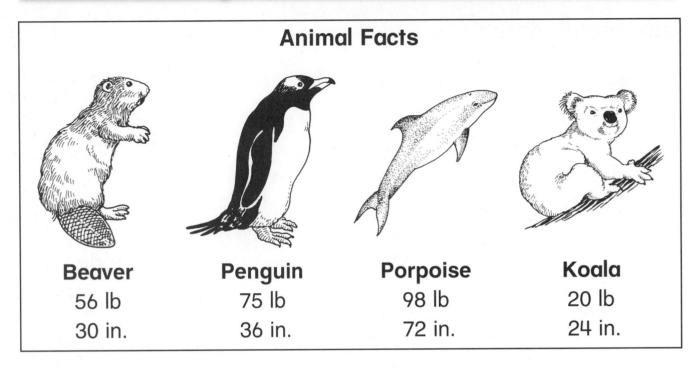

Animal Facts

Beaver	Penguin	Porpoise	Koala
56 lb	75 lb	98 lb	20 lb
30 in.	36 in.	72 in.	24 in.

Solve each problem.

1. Which is longer, the beaver or the penguin?

   _____

   How much longer? _____ inches longer

2. Which weighs more, a beaver and a koala together or a penguin?

   _____

3. Which weighs more, 2 beavers or a porpoise?

   _____

4. How much do a beaver and a koala weigh together?

   _____ pounds

Practice Set 45

1. Match the turn-around facts.

4 + 3 = 7   5 + 4 = 9

6 + 4 = 10   3 + 4 = 7

4 + 5 = 9   4 + 6 = 10

Write the turn-around fact.

2. 8 + 1 = 9

_____ + _____ = _____

3. 3 + 5 = 8

_____ + _____ = _____

4. 2 + 9 = 11

_____ + _____ = _____

5. 6 + 1 = 7

_____ + _____ = _____

---

6. How many pennies make 1 dime? Circle s.

Write the number that is 10 more.

Example 15, _25_

7. 26, _____

8. 51, _____

9. 69, _____

10. 82, _____

11. 9, _____

Practice Set 46

 Add. Remember to practice your addition facts.

1. $3 + 0 =$ _____

2. $4 + 0 =$ _____

3. $5 + 0 =$ _____

4. $6 + 0 =$ _____

5. $7 + 0 =$ _____

6. $8 + 0 =$ _____

7. $4 + 1 =$ _____

8. $5 + 1 =$ _____

9. $6 + 1 =$ _____

10. $7 + 1 =$ _____

11. $8 + 1 =$ _____

12. $9 + 1 =$ _____

Answer each question.

13. How many children are 44 inches tall? _____ children

14. How many children are 46 inches tall? _____ children

15. How many more children are 44 inches tall than 46 inches tall?

_____ more children

Heights of First Grade	
Inches	**Tallies**
42	̶H̶H̶T̶
43	̶H̶H̶T̶ ̶H̶H̶T̶ ̶H̶H̶T̶ /
44	̶H̶H̶T̶ ̶H̶H̶T̶ /
45	̶H̶H̶T̶ //
46	////
47	//

Practice Set 47

Write the missing rules and numbers.

1.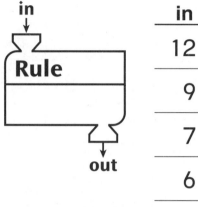

in	out
12	9
9	6
7	4
6	
Your turn:	

2.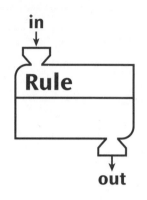

in	out
6	10
5	9
11	15
15	
Your turn:	

3.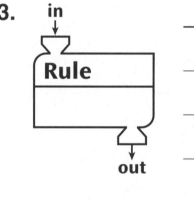

in	out
2	12
8	18
7	17
5	
Your turn:	

4.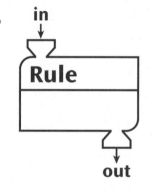

in	out
4	8
5	10
6	12
7	
Your turn:	

5. 🖉 **Writing/Reasoning** How did you find the rule in Problem 4? Explain your answer.

_____

_____

Practice Set 48

MRB
6 96
100–102

Follow the rule. Write the *out* numbers.

1.

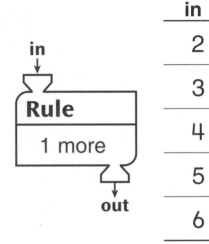

in

Rule

1 more

out

in	out
2	3
3	4
4	
5	
6	

2.

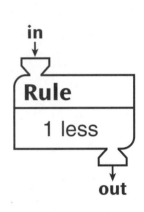

in

Rule

1 less

out

in	out
7	6
8	
9	
10	
11	

Circle *even* or *odd*.

3.

even   odd

4.

even   odd

5.

even   odd

6.

even   odd

Fill in the missing numbers. Count back.

7. 29, ____, ____, 26, ____, ____, ____, ____, ____, ____

Use with or after Lesson 5•13.

Practice Set 49

MRB
6 7 16
100

Circle the names that belong in each box.

1.

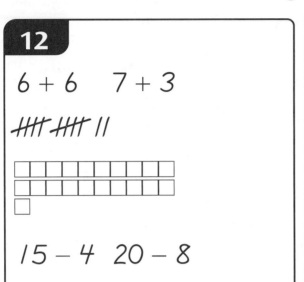

```
12
6 + 6    7 + 3

//// //// //

15 – 4   20 – 8
```

2.

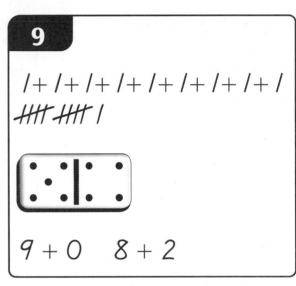

```
9
/ + / + / + / + / + / + / + / + /
//// //// /

9 + 0    8 + 2
```

FACTS PRACTICE **Add. Remember to practice your addition facts.**

3. 1 + 1 = ____   **4.** 2 + 2 = ____   **5.** 3 + 3 = ____

6. 4 + 4 = ____   **7.** 5 + 5 = ____   **8.** 6 + 6 = ____

9. 7 + 7 = ____   **10.** 8 + 8 = ____   **11.** 9 + 9 = ____

Follow the rule. Write the *out* numbers.

12.

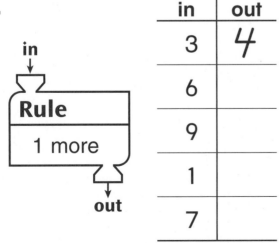

in	out
3	4
6	
9	
1	
7	

Rule: 1 more

13.

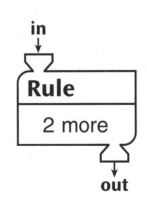

in	out
4	6
5	
6	
7	
8	

Rule: 2 more

Practice Set 50

Write 3 numbers for each domino.

1.

_____ , _____ , _____

2.

_____ , _____ , _____

3.

_____ , _____ , _____

4.

_____ , _____ , _____

5.

_____ , _____ , _____

6.

_____ , _____ , _____

Estimate. Write *yes* or *no*.

7. Will $18 + 5$ be more than 20? _____

8. Will $20 - 7$ be less than 10? _____

9. Will $15 + 2$ be less than 20? _____

10. Will $20 - 2$ be more than 15? _____

11. Will $17 - 5$ be more than 10? _____

12. Will $10 + 8$ be more than 20? _____

 Add. Remember to practice your addition facts.

13. $7 + 3 =$ _____

14. $2 + 8 =$ _____

15. $4 + 6 =$ _____

16. $9 + 1 =$ _____

Use with or after Lesson 6·3.

Practice Set 51

Write the fact family for the Fact Triangle.

Example

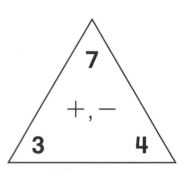

$7 = \underline{3} + \underline{4}$

$7 = \underline{4} + \underline{3}$

$7 - \underline{3} = \underline{4}$

$7 - \underline{4} = \underline{3}$

1.

$5 = \underline{\hspace{1cm}} + \underline{\hspace{1cm}}$

$5 = \underline{\hspace{1cm}} + \underline{\hspace{1cm}}$

$5 - \underline{\hspace{1cm}} = \underline{\hspace{1cm}}$

$5 - \underline{\hspace{1cm}} = \underline{\hspace{1cm}}$

How much money?

2.

$0.\underline{\hspace{1cm}}$ or $\underline{\hspace{1cm}}$¢

3.

$0.\underline{\hspace{1cm}}$ or $\underline{\hspace{1cm}}$¢

 Add. Remember to practice your addition facts.

4. $4 + 2 = \underline{\hspace{1.5cm}}$

5. $3 + 1 = \underline{\hspace{1.5cm}}$

6. $8 + 1 = \underline{\hspace{1.5cm}}$

7. $5 + 4 = \underline{\hspace{1.5cm}}$

Practice Set 52

MRB
7 66

Measure each line segment. Use centimeters.

1.

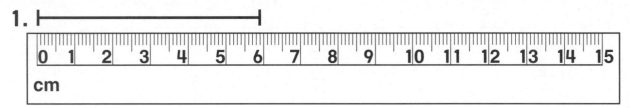

about _____ centimeters

2.

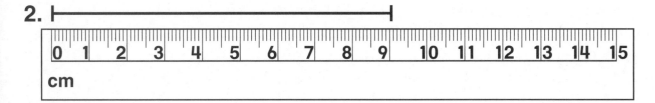

about _____ centimeters

3.

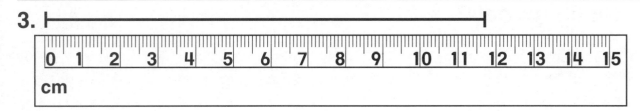

about _____ centimeters

Write the missing numbers.

	Before	Number	After
Example	49	50	51
4.		32	
5.	21		23

	Before	Number	After
6.		14	15
7.		60	
8.	39	40	

Use with or after Lesson 6•6.

Practice Set 53

Find the sum. Then circle *even* or *odd*.

1. $2 + 3 =$ _____   **even   odd**

2. $4 + 5 =$ _____   **even   odd**

3. $10 + 10 =$ _____   **even   odd**

4. $11 + 3 =$ _____   **even   odd**

5. $100 + 1 =$ _____   **even   odd**

6. $\begin{array}{r} 7 \\ + \ 3 \\ \hline \end{array}$ **even   odd**

7. $\begin{array}{r} 7 \\ + \ 4 \\ \hline \end{array}$ **even   odd**

8. $\begin{array}{r} 9 \\ + \ 2 \\ \hline \end{array}$ **even   odd**

9. **Writing/Reasoning** How do you know if a number is even or odd? Explain your answer by using words, numbers, or pictures.

Name _____ Date _____ Time _____

Practice Set 54

Follow the rule. Complete the table.

1.

Rule
Add 5

in	out
5	
10	
25	
15	
30	

2.

Rule
Add 2

in	out
	7
	25
	16
	73
	39

What time is it?

3.

quarter-to _____

4.

half-past _____

Write another name for the number.

5. 15 _____

6. 27 _____

7. 32 _____

8. 50 _____

Use with or after Lesson 6•8.

Practice Set 55

How much money?

1.

$0._____

2.

_____¢

3.

_____¢

4.

$0._____

5.

_____¢

6.

$0._____

Use your calculator. Count by 5s.

Press (ON/C) (5) (+) (5) (=) (=) (=) (=)

7. __5__, __10__, __15__, __20__, ____, ____, ____, ____, ____,

____, ____, ____, ____, ____, ____, ____, ____,

____, ____, ____, ____, ____, ____, ____,

Name _____ Date _____ Time _____

Practice Set 56

Match the clocks.

Example

Write 3 numbers for each domino.

4.

_____ , _____ , _____

1.

5.

_____ , _____ , _____

6.

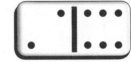

_____ , _____ , _____

2.

7.

3.

8.

_____ , _____ , _____

70

Use with or after Lesson 6·10.

Practice Set 57

Use the pictures to answer the problems.

1. Are you more likely to spin a shaded part or a white part? _____

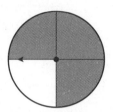

2. Are you more likely to spin a shaded part or a white part? _____

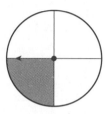

3. **Writing/Reasoning** Explain how you found the answer to Problem 2.

_____

_____

Draw the hour hand and the minute hand.

4.

3:00

5.

11:15

Add. Remember to practice your addition facts.

6. 7 + 2 = _____

7. 3 + 6 = _____

8. 1 + 5 = _____

9. 0 + 10 = _____

Practice Set 58

The children are in first grade.

They are in Ms. Lund's reading group.

They are this tall:

| Sally 41 in. | Kendra 46 in. | Drew 46 in. | Ken 46 in. | Lorna 49 in. |

1. What is the **difference** between the largest number and the smallest number?

    49
    − 41
    ――――

2. What is the **range**? _____

3. About how tall are first graders?

    They are about _____ inches tall.

Fill in the five-minute marks.

4.

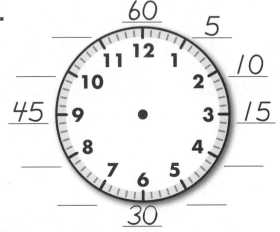

5.

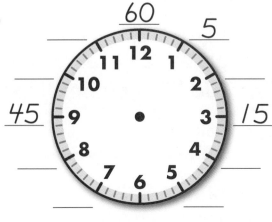

Practice Set 59

1. Circle the triangles.

2. Circle the squares.

Fill in the rule.

Example

Rule
Count down by 1s

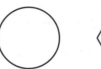 17   16   15   14   13

3.

Rule

 25   27   29   31   33

4.

Rule

 10   13   16   19   22

5. 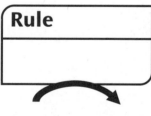 **Writing/Reasoning** Explain how you found the rule in Problem 4.

Practice Set 60

1. Color the smallest rectangle.

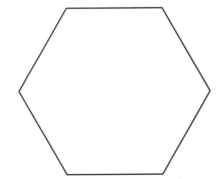

---

2. Color the largest hexagon.

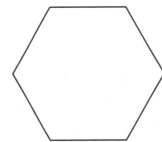

List the fact family for each Fact Triangle.

Example

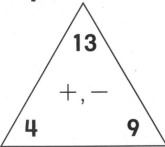

$4 + 9 = 13$
$9 + 4 = 13$
$13 - 4 = 9$
$13 - 9 = 4$

3.

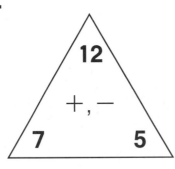

_____
_____
_____
_____

4.

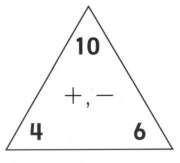

_____
_____
_____
_____

Practice Set 61

Color the one that is the same size and shape.

1.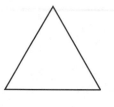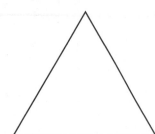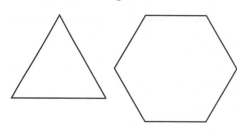

2.

What time is it? Circle the correct time.

3.

5:40

6:40

6:20

4.

3:15

2:15

2:45

5.

12:05

12:10

12:15

6.

7:25

7:35

7:45

Practice Set 62

1. Color all of the 4-sided polygons.

Complete the number family.

2.

3 Family	
0	3

3.

5 Family	
2	3

Estimate.

4. Will 17 − 5 be *more* or *less* than 10? _____

5. Will 7 + 5 be *more* or *less* than 10? _____

Practice Set 63

1. Circle the one that will roll.

How many flat faces? How many corners?

2. _____ flat faces

_____ corners

3. 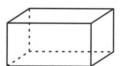 _____ flat faces

_____ corners

---

4. Circle all that are true.

$3 + 7 = 7 + 3$ $6 - 5 = 5 - 6$

$4 = 2 + 3$ $3 + 1 = 4$

Write 3 numbers for each domino.

5.

_____ , _____ , _____

6.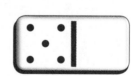

_____ , _____ , _____

7.

_____ , _____ , _____

---

8. **Writing/Reasoning** Write three number
sentences using the numbers 4, 3, and 7.

_____

_____

_____

Practice Set 64

Draw a line to the shape you can make using the ink pad.

Example

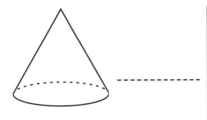

1.

2.

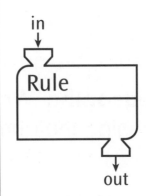

Fill in the rule box and the missing numbers.

3.

in	out
10	8
6	4
	10
14	
	7

4.

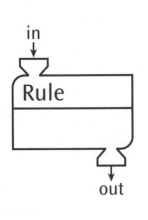

in	out
3	7
7	11
5	
	10
	12

Use with or after Lesson 7•6.

Practice Set 65

What will the shape look like after you cut it out and open it up? Draw a line to match.

1.

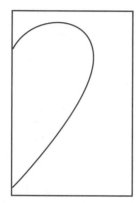

2.

3. Show 35 cents two different ways.

Practice Set 66

How many?

1. 25  = _____

2. 2  = _____

3. _____  = 1

4. _____  = 15

Draw hands to show the time.

5.

8:25

6.

5:45

7.

12:10

Write the missing numbers.

8. __98__, _____, __100__, _____, _____, _____, _____, _____

9. __112__, _____, _____, __115__, _____, _____, _____, _____

10. __55__, _____, _____, __58__, _____, _____, _____, _____

Use with or after Lesson 8·1.

Practice Set 67

Write the amount.

Example one dollar and sixty-two cents ___$1.62___

1. three dollars and fourteen cents _____

2. two dollars and seven cents _____

3. one dollar and ninety-one cents _____

4. eighty-nine cents _____

Write the missing number.

5. ____ pennies = 1 dollar

6. 1 dollar = ____ dimes

7. 1 dollar = ____ quarters

8. 20 nickels = ____ dollar

9. 1 dollar = ____ quarters and ____ dimes

Complete the number family.

10.

4 Family	
4	0

11.

5 Family	
2	3

12.

6 Family	
3	3

Practice Set 68

What number do the blocks show?

1.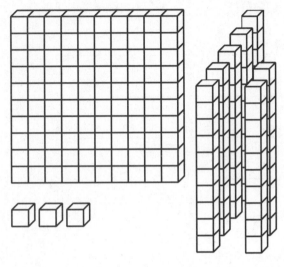

_____

2.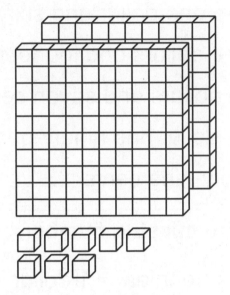

_____

Match the halves.

3.

4.

5.

Add. Remember to practice your addition facts.

6. $1 + 2 =$ _____   **7.** $3 + 2 =$ _____   **8.** $4 + 5 =$ _____

Use with or after Lesson 8·3.

Practice Set 69

Toy Store

ball	ring	crayons	bear
$0.20	12¢	$0.25	16¢

Solve each number story.

1. How much do 2 rings cost? $0. _____ or _____¢

2. How much do a bear and a ball cost? $0. _____ or _____¢

3. Which costs more, crayons or a ball? _____

   How much more? $0. _____ or _____¢ more

4. How much do crayons and a bear cost?

   $0. _____ or _____¢

5. **Writing/Reasoning** Write a number story about the Toy Store. Write a number model for your story.

   _____

   _____

   _____

   _____

Practice Set 70

How much change?

1.

16¢

_____

2.

25¢

_____

Write the addition turn-around facts.

Example

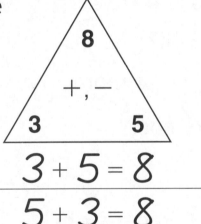

8

+, −

3   5

$3 + 5 = 8$

$5 + 3 = 8$

3.

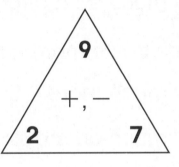

9

+, −

2   7

_____

_____

4.

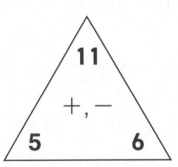

11

+, −

5   6

_____

_____

5.

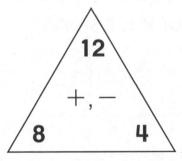

12

+, −

8   4

_____

_____

Use with or after Lesson 8•5.

Practice Set 71

Divide into equal parts.

1. Show three different ways to divide
   a rectangle in half.

2. Show three different ways to divide
   a rectangle into fourths.

How many equal parts does each shape have?

3. 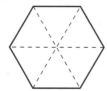 _____

4. 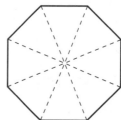 _____

5. **Writing/Reasoning** Is this cracker divided
   into equal parts? Explain your answer by
   using words, pictures, or numbers.

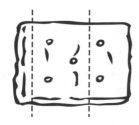

Practice Set 72

Write the fraction for the shaded part.

Example

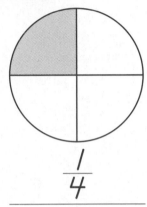

$\dfrac{1}{4}$

1.

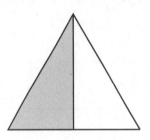

_____

2.

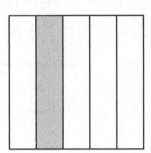

_____

3.

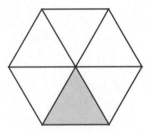

_____

4.

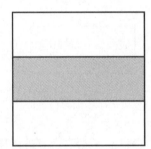

_____

5.

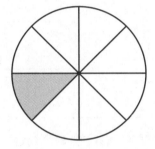

_____

FACTS PRACTICE   **Add. Remember to practice your addition facts.**

6. _____ = 5 + 6   **7.** 7 + 3 = _____   **8.** 9 + 9 = _____

9. _____ = 6 + 9   **10.** _____ = 4 + 5   **11.** 2 + 9 = _____

12.   7
    + 7

13.   8
    + 4

14.   9
    + 3

15.   2
    + 8

16.   4
    + 9

Practice Set 73

Divide into equal shares. Circle the number of pennies each child will get.

Example   Marcus and Albert

1. Sandy and Brenda

2. Tim, Tom, and Ted

Write the 3-digit number.

3. 4 in the hundreds place

7 in the tens place

2 in the ones place

_____

4. 9 in the hundreds place

1 in the tens place

6 in the ones place

_____

5. 3 in the hundreds place

8 in the tens place

5 in the ones place

_____

6. 2 in the hundreds place

0 in the tens place

4 in the ones place

_____

Practice Set 74

1. Count by 10s. Start at 6.
   Write an **X** over each number you count.

									0
1	2	3	4	5	X̶6	7	8	9	10
11	12	13	14	15	X̶16	17	18	19	20
21	22	23	24	25	26	27	28	29	30
31	32	33	34	35	36	37	38	39	40
41	42	43	44	45	46	47	48	49	50
51	52	53	54	55	56	57	58	59	60
61	62	63	64	65	66	67	68	69	70
71	72	73	74	75	76	77	78	79	80
81	82	83	84	85	86	87	88	89	90
91	92	93	94	95	96	97	98	99	100
101	102	103	104	105	106	107	108	109	110
111	112	113	114	115	116	117	118	119	120

Fill in the blanks.

Example 215 = __2__ hundreds __1__ tens __5__ ones

2. 87 = _____ tens _____ ones

3. 394 = _____ hundreds _____ tens _____ ones

4. 762 = _____ hundreds _____ tens _____ ones

5. 251 = _____ hundreds _____ tens _____ ones

88   Use with or after Lesson 9·1.

Practice Set 75

Write the missing numbers.
Use your number grid.

1.

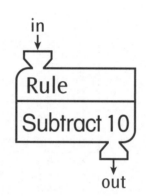

in
↓
Rule
Add 10
↓
out

in	out
60	
83	
47	
162	
215	

2.

in
↓
Rule
Subtract 10
↓
out

in	out
90	
76	
102	
210	
238	

Add. Remember to practice your addition facts.

3.   5
     + 3
    ____

4.   6
     + 4
    ____

5.   3
     + 2
    ____

6.   5
     + 0
    ____

7.   7
     + 1
    ____

8.   8
     + 5
    ____

9.   9
     + 1
    ____

10.   3
      + 6
     ____

11.   4
      + 7
     ____

12.   4
      + 2
     ____

13. ____ = 6 + 6   **14.** 3 + 9 = ____   **15.** 8 + 3 = ____

Practice Set 76

Fill in each number-grid piece below.

Example

14
24
34
44
54
64
74

−9	−8	−7	−6	−5	−4	−3	−2	−1	0
1	2	3	4	5	6	7	8	9	10
11	12	13	14	15	16	17	18	19	20
21	22	23	24	25	26	27	28	29	30
31	32	33	34	35	36	37	38	39	40
41	42	43	44	45	46	47	48	49	50
51	52	53	54	55	56	57	58	59	60
61	62	63	64	65	66	67	68	69	70
71	72	73	74	75	76	77	78	79	80
81	82	83	84	85	86	87	88	89	90
91	92	93	94	95	96	97	98	99	100
101	102	103	104	105	106	107	108	109	110

1. [39] [ ] ... [ ] [80]

2. 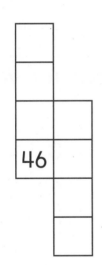 [46]

3. 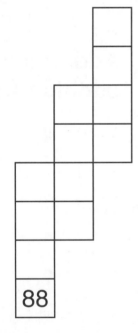 [88]

4. ✏️ **Writing/Reasoning** Will 8 + 6 be more
or less than 10? Explain your answer.

_____

_____

_____

Practice Set 77

MRB
8
64 71

Animal Facts

Fish	**Woodpecker**	**Black Bear**	**Fox**
15 oz	2 oz	300 lb	14 lb
12 in.	8 in.	60 in.	20 in.

Solve each problem.

1. Which is shorter, the fox or the fish? _____

   How much shorter? _____ inches shorter

2. Which weighs less, the woodpecker or the fish?

   _____

   How much less? _____ ounces less

3. What is the total length of the fox and the black bear?

   Their total length is _____ inches.

4. What is the total weight of the fish and the woodpecker?

   Their total weight is _____ ounces.

Practice Set 78

Complete each design so that the two halves match.

Example

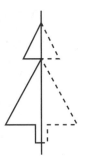

1.

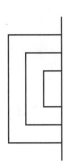

2.

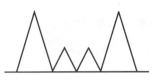

3.

4.

5.

6. ✏️ **Writing/Reasoning** Write a number story about the picture below.

7 ft.

4 ft.

_____

_____

_____

Solve.

7. 26 − 9 = _____   **8.** 36 − 9 = _____   **9.** 96 − 9 = _____

10. 46 + 9 = _____   **11.** 82 + 5 = _____   **12.** 32 + 5 = _____

    Use with or after Lesson 9•5.

Practice Set 79

1. Color one-half of the circle.

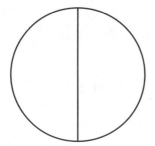

2. Color three-fourths of the square.

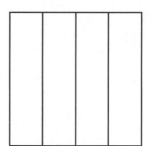

Fill in the rule box. Complete the frames.

3.
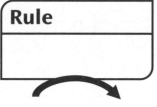
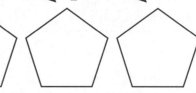

210   200   190

4.
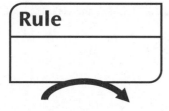
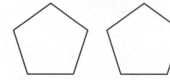

$0.37 $0.47 $0.57

5.
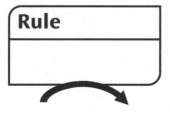
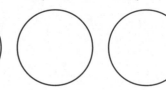

26   30   34

6.
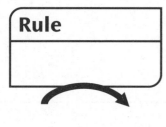
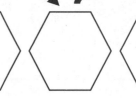

15   18

Practice Set 80

Write the fraction for the shaded parts.

1.

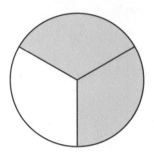

☐ parts that are shaded
——
☐ number of equal parts

2.

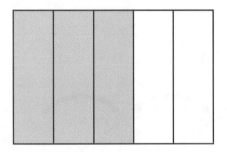

☐ parts that are shaded
——
☐ number of equal parts

Write 5 more names for each number.

3.

25
10 + 5 + 5 + 5
30 − 5

4.

40
38 + 2
50 − 10

Use with or after Lesson 9•7.

Practice Set 81

1. Write the fraction.

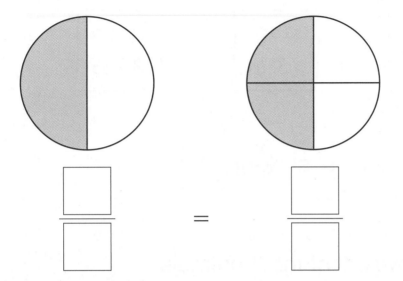

$$\boxed{} \over \boxed{} = \boxed{} \over \boxed{}$$

Write the number.

Example four hundred fifteen __415__

2. ninety-two _____

3. seven hundred eighty-three _____

4. two hundred seven _____

Use the three digits. Write the smallest and largest numbers.

Example 2, 8, 5   smallest __258__   largest __852__

5. 7, 9, 3   smallest _____   largest _____

6. 9, 1, 4   smallest _____   largest _____

7. 6, 8, 5   smallest _____   largest _____

Practice Set 82

Animal Weights

fox	goose	dog	koala
14 lb	18 lb	18 lb	20 lb

1. Which weight is on the most stick-on notes?

    _____ pounds

2. What is the **typical** weight of the 4 animals?

    _____ pounds

3. What is the **middle** weight of the 4 animals?

    _____ pounds

4. Complete the grid.

231	232	233	234	235			238	239	240
241				245		247		249	
		253					258		260
261					266				
271		273	274			277			280
		283					288	289	
									300

Practice Set 83

Complete the shapes so that the two halves match.

1.

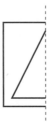

2.

Use <, >, or =.

< means *is less than*
> means *is greater than*
= means *is equal to*

3. 2 dimes ☐ 15¢

4. 25¢ ☐ 6 nickels

5. 49¢ ☐ $0.49

6. 1 quarter, 1 dime, 2 pennies ☐ $0.38

7. 52¢ ☐ 2 quarters, 2 pennies

8. **Writing/Reasoning** Explain how you got your answer to Problem 7 using words, numbers, or pictures.

Practice Set 84

Mark the coins you need.

1.

74¢

2.

49¢

3.

85¢

Addition Patterns	Subtraction Patterns
4. $4 + 10 =$ _____	**9.** _____ $= 62 - 10$
5. $4 + 20 =$ _____	**10.** _____ $= 62 - 20$
6. $4 + 30 =$ _____	**11.** _____ $= 62 - 30$
7. $4 + 40 =$ _____	**12.** _____ $= 62 - 40$
8. $4 + 50 =$ _____	**13.** _____ $= 62 - 50$

Use with or after Lesson 10•3.

Practice Set 85

Museum Store

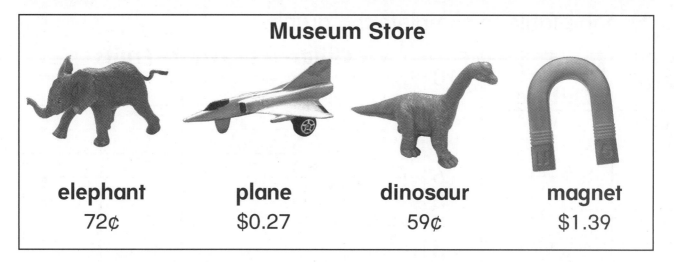

elephant	plane	dinosaur	magnet
72¢	$0.27	59¢	$1.39

Solve each problem.

1. Which costs more, the elephant or the dinosaur?

   _____

   How much more? _____¢ more

2. Which costs less, the dinosaur or the plane?

   _____

   How much less? $0._____ less

3. Which costs more, 2 dinosaurs or a magnet?

   _____

   How much more? _____¢ more

4. If you paid for an elephant with 3 quarters,
   how much change would you get?

   I would get _____¢ change.

Practice Set 86

1. Use the table to complete the graph.

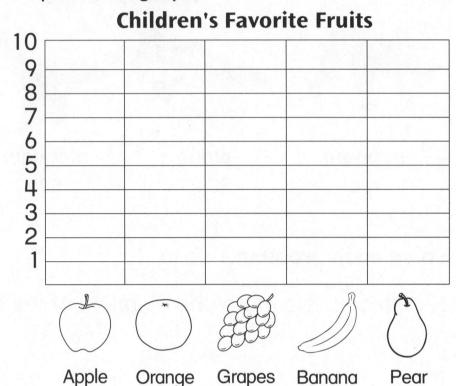

Fruit	Number of Children
apple	4
orange	3
grapes	3
banana	6
pear	2

Children's Favorite Fruits

Apple   Orange   Grapes   Banana   Pear

2. Which fruit do children like best? _____

3. Which fruit do children like least? _____

Mark the line segment to show each length.

4. 8 centimeters

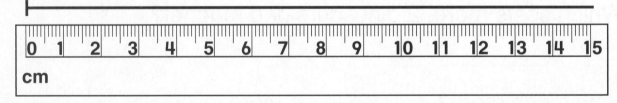

5. 14 centimeters

Use with or after Lesson 10·5.

Practice Set 87

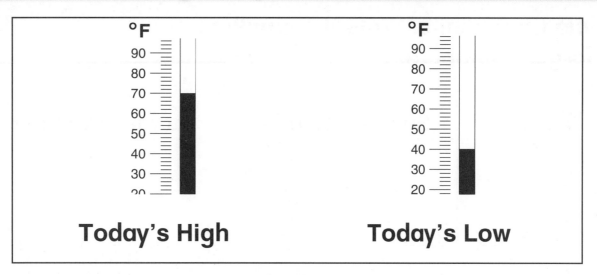

Today's High   **Today's Low**

1. Find the difference between
the two temperatures.   _____°F

**How likely are you to pick each shape
without looking? Write *certain*, *likely*,
unlikely, or *impossible*.**

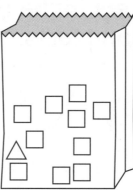

2. ⬤   **3.** ☐

_____   _____

4. △   **5.** ⇨

_____   _____

Practice Set 88

MRB
10 27

Use the three digits. Write the smallest and largest numbers.

Digits	Smallest Number	Largest Number
3, 2, 7	**1.**	**2.**
5, 4, 8	**3.**	**4.**
2, 7, 9	**5.**	**6.**
6, 4, 6	**7.**	**8.**
1, 5, 3	**9.**	**10.**

Write the missing numbers in the Fact Triangle. Then finish the turn-around facts.

11.

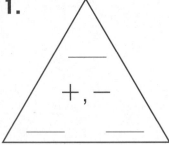

$$\_\_\_ + 5 = 7$$

$$7 = 5 + \_\_\_$$

12.

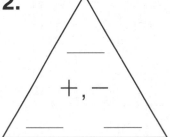

$$\_\_\_ = 8 + 3$$

$$3 + 8 = \_\_\_$$

13.

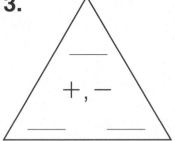

$$4 + \_\_\_ = 10$$

$$10 = \_\_\_ + 4$$

Test Practice ⟨1⟩

Fill in the circle next to your answer.

1. Julio wants to find out how warm it is outside. What should he use?

Ⓐ measuring cup

Ⓑ thermometer

Ⓒ  bathroom scale

Ⓓ 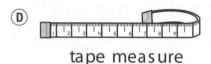 tape measure

2. Shen's family searched for shells at the beach. Shen found 3 shells, her dad found 6 shells, and her mom found 9 shells.

How many 🐚 should Shen draw in the last row?

SHELLS WE FOUND	
Shen	🐚
Dad	🐚 🐚
Mom	

🐚 = 3

Ⓐ 1   Ⓑ 2   Ⓒ 3   Ⓓ 9

3. Cam put 10 blocks in a jar. Her sister picks a block without looking. Which block is she **most likely** to pick?

Ⓐ

Ⓑ

Ⓒ

Ⓓ

Test Practice 〈**1**〉 *continued*

Fill in the circle next to your answer.

4. Jamal has 9 pictures. Kate has 5 pictures. About how many pictures do they have all together?

ⓐ less than 1   ⓑ less than 5

ⓒ less than 10   ⓓ less than 15

5. Kit used a book to measure a jump rope.

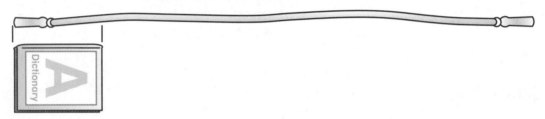

About how many books long is Kit's jump rope?

ⓐ 2   ⓑ 4   ⓒ 6   ⓓ 9

6. Look at the number pattern that Jack made.

0, 5, 10, 15, 20, 25

What is his rule?

ⓐ Skip count by 5s.   ⓑ Count back by 1s.

ⓒ Add even numbers.   ⓓ Subtract by 3s.

7. Mulan is making a calendar. Which number will she write on Tuesday?

ⓐ 9   ⓑ 10

ⓒ 11   ⓓ 12

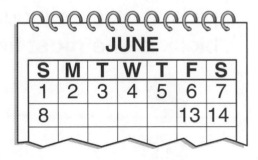

JUNE						
S	M	T	W	T	F	S
1	2	3	4	5	6	7
8					13	14

Test Practice 2

Fill in the circle next to your answer.

1. Which fish is the **longest**?

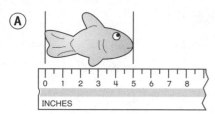

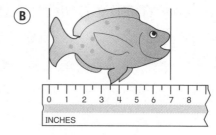

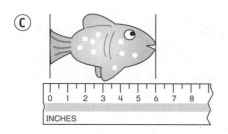

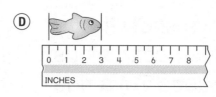

2. Which number belongs in the box?

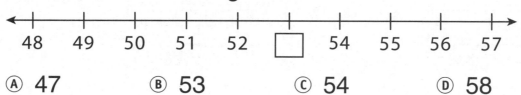

Ⓐ 47   Ⓑ 53   Ⓒ 54   Ⓓ 58

3. Which sign belongs in the box?

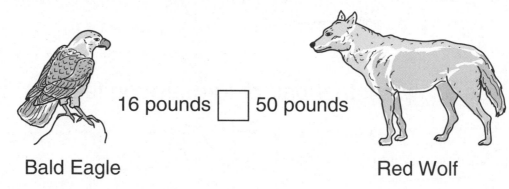

16 pounds ☐ 50 pounds

Bald Eagle   Red Wolf

Ⓐ >   Ⓑ $ Ⓒ =   Ⓓ <

Test Practice ◆ 2 ▶ *continued*

Fill in the circle next to your answer.

4. Mrs. Bhatia fed 17 chicks. Then 4 of them fell asleep. How many chicks are awake?

Ⓐ 3   Ⓑ 4   Ⓒ 13   Ⓓ 21

5. Toya's blocks show the number on her school bus. Which bus does Toya ride?

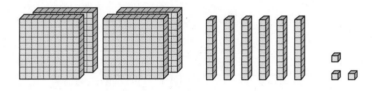

Ⓐ
364

Ⓑ
436

Ⓒ
463

Ⓓ
643

6. The cards below show a pair of turn-around facts.

$7 + 5 = 12$ $5 + 7 = 12$

Which cards show another pair of turn-around facts?

Ⓐ
$10 - 3 = 7$ $5 + 2 = 7$

Ⓑ
$9 + 3 = 12$ $4 + 8 = 12$

Ⓒ
$11 - 6 = 5$ $8 - 3 = 5$

Ⓓ
$6 + 8 = 14$ $8 + 6 = 14$

Test Practice 3

Fill in the circle next to your answer.

1. Which fact belongs in the same fact family
   as 13 − **4** = **9**?

   Ⓐ 13 + 4 = 17   Ⓑ 13 − 9 = 4

   Ⓒ 9 − 4 = 5   Ⓓ 9 + 9 = 18

2. Leona saw 11 birds at the beach.
   Which shows how many birds she saw?

   Ⓐ //// ////   Ⓑ //// //// //

   Ⓒ //// /   Ⓓ //// //// /

3. Tyrone and Selma found some pennies.

   Selma's pennies

   Tyrone's pennies

   How many **more** pennies did Selma find?

   Ⓐ 21   Ⓑ 8   Ⓒ 5   Ⓓ 4

Test Practice ◆ 3 ◆ *continued*

Fill in the circle next to your answer.

4. Ann used four of these triangles to make a shape.

Which shape did Ann make?

Ⓐ 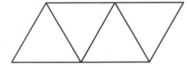   Ⓑ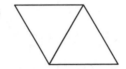

Ⓒ 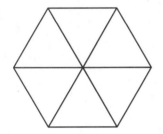   Ⓓ

5. Which shape has six sides?

Ⓐ 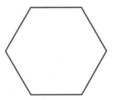   Ⓑ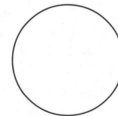

Ⓒ 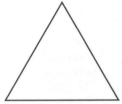   Ⓓ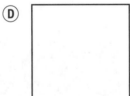

6. Which number belongs in this box? $9 + \boxed{} = 17$

    Ⓐ 1   Ⓑ 7   Ⓒ 8   Ⓓ 26

Test Practice 4

Fill in the circle next to your answer.

1. Mick and Ella put away their toys. Mick's box holds about 20 toys. Ella's box holds fewer toys. **About** how many toys can fit in Ella's box?

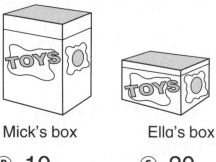

Mick's box   Ella's box

Ⓐ 5   Ⓑ 10   Ⓒ 20   Ⓓ 40

2. Mei feeds her dog at 7 o'clock each day. Which clock shows **about** 7 o'clock?

Ⓐ    Ⓑ

Ⓒ   Ⓓ

3. Which card is equal to this card?

12

Ⓐ 8 + 5   Ⓑ 15 − 3   Ⓒ 13 − 2   Ⓓ 6 + 4

Test Practice 4   *continued*

Fill in the circle next to your answer.

4. Zoe and Kit count their pennies. Zoe has 29 pennies. Kit has 16 pennies. Which number sentence shows how many more pennies Zoe has than Kit?

Ⓐ $29 - 16 = 13$ Ⓑ $13 + 3 = 16$

Ⓒ $29 - 13 = 16$ Ⓓ $29 + 16 = 45$

5. Berto saw 11 ladybugs. Then 5 flew away. How many ladybugs were left?

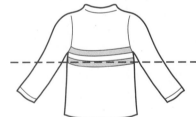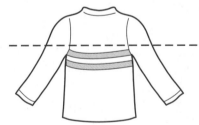

Ⓐ 4   Ⓑ 6   Ⓒ 8   Ⓓ 22

6. Don folds his shirt. He makes both halves match. Which picture shows how Don folds his shirt?

Ⓐ   Ⓑ

Ⓒ   Ⓓ

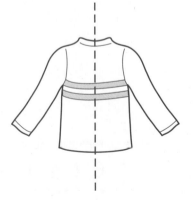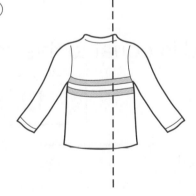

                                                Use with or after Unit 10.